AF575656

LEGENDS OF WARFARE

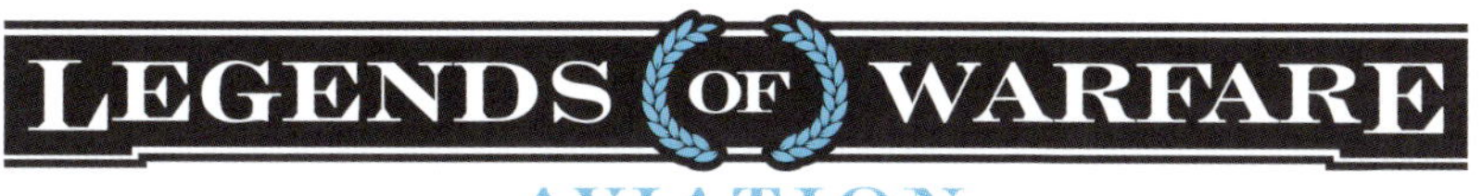

AVIATION

F-111 Aardvark

General Dynamics' Variable-Swept-Wing Attack Aircraft

JOHN GOURLEY

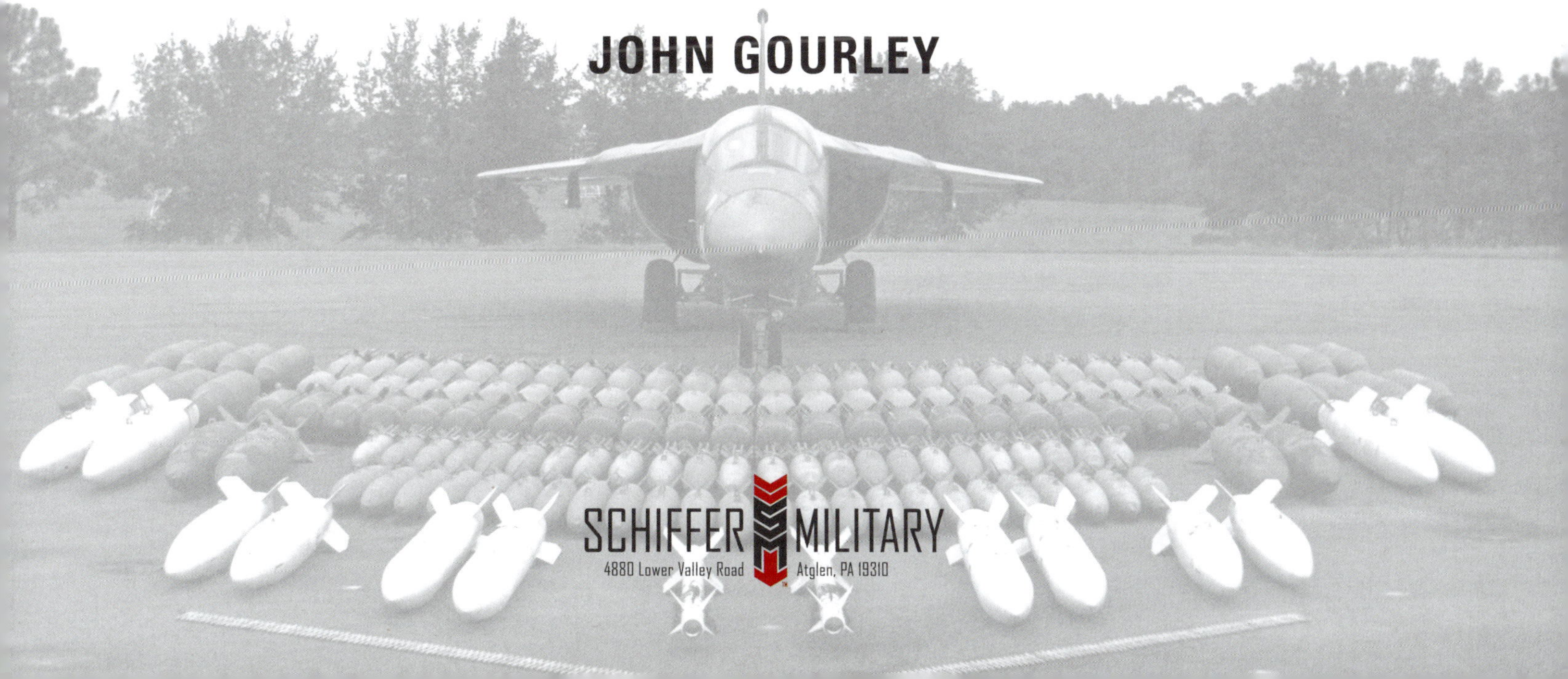

SCHIFFER MILITARY
4880 Lower Valley Road Atglen, PA 19310

Library of Congress Control Number: 2020943402

Designed by Justin Watkinson
Type set in Impact/Minion Pro/Univers LT Std

ISBN: 978-0-7643-6128-9
Printed in China

Published by Schiffer Publishing, Ltd.
4880 Lower Valley Road
Atglen, PA 19310
Phone: (610) 593-1777; Fax: (610) 593-2002
E-mail: Info@schifferbooks.com
www.schifferbooks.com

For our complete selection of fine books on this and related subjects, please visit our website at www.schifferbooks.com. You may also write for a free catalog.

Schiffer Publishing's titles are available at special discounts for bulk purchases for sales promotions or premiums. Special editions, including personalized covers, corporate imprints, and excerpts, can be created in large quantities for special needs. For more information, contact the publisher.

We are always looking for people to write books on new and related subjects. If you have an idea for a book, please contact us at proposals@schifferbooks.com.

Dedication

To everyone who ever was involved in the design, production, and documentation; flew; maintained; or supported the F-111. Its ability to fly long distances unrefueled; to use the Earth's contours to fly low, fast, and unseen; to be able to attack targets in all types of weather; and to use adjustable swept wings and turbofan engines was truly an aviation marvel in many respects. The F-111 was a major factor in the Cold War remaining cold, the best outcome any weapon designer can achieve.

Acknowledgments

The author would like to recognize and thank the following individuals who have contributed to, advised on, provided photos for, and showed an interest as this book was in its many phases of preparation. Dave Baker and Phillip Blaha have always had an interest in my work, and for that you have my eternal gratitude. Lou Drendel got me started in book writing, and it has been a pleasure of my time ever since. Dennis Jenkins provided many of the historical images this book contains. Dr. Carlo Kopp, Paul Minert, Norm Taylor, and Kurt Todoroff were always helpful and provided informative data as the book was being created. To the firefighters of "A-Shift" at Eglin AFB, since it was we who were on duty on October 16, 1976, when F-111E 67-0116 crashed. Our textbook efforts in crash/rescue tactics that day I will always remember, as well as the day I took part in the postcrash activities.

To my daughter, Jacqueline, who has always been a personal source of positive energy in my life experience and who had to patiently put up with me on extended photo-taking and information-gathering outings in places as far flung as White Sands, New Mexico, and the National Museum of Naval Aviation in Pensacola, Florida. To my sister Lauren, my nephew Brandon, and niece Marissa, who have contributed to this in sentimental ways only family members can relate to and appreciate.

Contents

F-111A under assembly at the General Dynamics facility in Ft. Worth, Texas. Platforms with safety rails provide access to wings, whether swept or fully forward. Each station has its own hydraulics, jacks, air pressure, electric power, and portable workstands to enable complete integration of components without interruption. Technicians of several specialties work together to fit, check out, and connect up all the parts that made up the F-111A. The fuselage structure consisted of aluminum, composites, and several grades of steel and titanium. *General Dynamics*

F-111A, 639766, towed across the parking ramp, with wings at the 16-degree, low-airspeed setting. During this early phase of flight operations, the plane has this light-gray/white undersides paint job. The radome is white with flat-black topsides to reduce sunlight glare for the aircrew. The nose pitot boom has instrumentation probes, used during the early flight tests to compare actual flight values with calculated estimates of expected readings. Compensating for the long nose, the tow bar has to be of exceptional length to give adequate clearance between the tug and the delicate nose boom. *General Dynamics*

CHAPTER 2

First Flights and the RF-111A

October 1964. F-111A, 63-9766, is displayed as Secretary of Defense Robert S. McNamara speaks before a crowd during the rollout ceremony. *Both photos from General Dynamics*

The first flight was successful. Wings were locked in position. High speeds were not attempted during this local maiden hop. The second flight, on January 6, 1965, would be more involved. The wings, which had been stationary at the 26-degree setting on flight #1, were moved forward to the 16-degree position, then moved back to the maximum 72.5 degrees. With this milestone accomplished (with the additional bonus of $875,000 to GD for demonstrating wing sweep far ahead of the contract stipulation), the decision was made to increase power to the TF30 engines to begin exploring the supersonic flight regime. As airspeed picked up, the flight plan suddenly went from cautiously optimistic to one of disappointment as both engines experienced compressor stalls, with this occurring in straight and level flight. With the second flight terminated prematurely, it was now obvious that time would be needed to remedy deficiencies related to air intake design. The ducts were shortened during the initial design phase, and with the layout at this stage for the most part frozen, no radical fuselage redesigns were an option. With flight testing suspended so inlet trouble shooting could become a priority task, and with a new engine variant, the TF30-P-3, installed and flown with no resolution of the stalling phenomenon, fixes were found by adding vortex generators to the intakes as well as adding adjustable air intake lips to counteract entry of unstable air into the engine's fan blades. With the newly modified "Triple Plow I" intakes attached, flights resumed in March. On the ninth flight, supersonic speeds were finally achieved, with a permanent fix not yet implemented. Maneuverability was restricted as the flight envelope was slowly, gradually expanded. Meanwhile, plans moved ahead for a new variant, the RF-111A, to be produced. The eleventh example was intended to show how the weapons bay could be adapted further to add a reconnaissance pallet, since it was already capable of accommodating a rotary cannon, bombs, missiles, and atomic weapons in any number of tactical roles. This would promote the F-111 even more as the multiuse platform it was claimed to be. Critical intelligence could be obtained by using the plane to sneak in and out of denied territory, taking pictures and using the infrared/radar packages to complete the haul of essential data. Such a pallet would take only an hour or less to install or remove. Again, reality took over as this adaptation flew for the first time in 1967, proving to be more complex than originally stated, and cost the program $118 million to fly as a prototype. In the end, none of the RF-111A examples were procured.

McNamara briefs the press after the rollout ceremony. He was a firm believer of the F-111 Air Force / Navy concept. *General Dynamics*

Gen. Irving R. Branch with F-111 test flight crewman Fred Voorhies, *center*, and Val Prahl, who flew many early developmental sorties. *General Dynamics*

US vice president Hubert H. Humphrey is given a guided tour of an F-111A at Carswell AFB, Texas. Political leaders followed the F-111's progress with great interest. *General Dynamics*

A press debrief is conducted after F-111 ship 1, flight 2, concluded. Media interest remained steady throughout the F-111's service introduction. *General Dynamics*

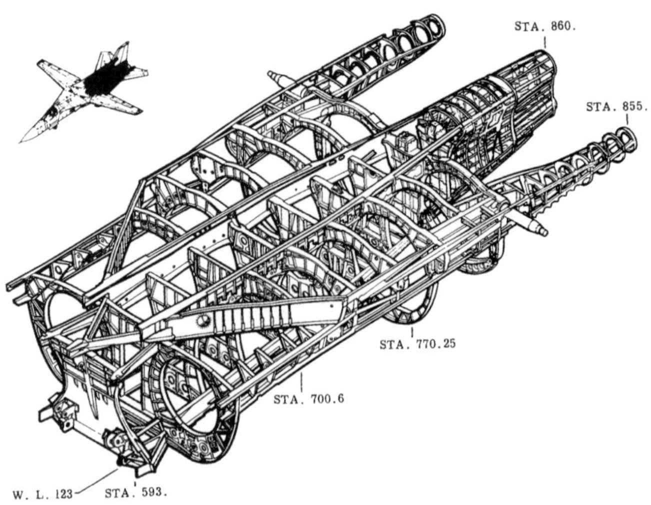

AFT FUSELAGE STRUCTURE

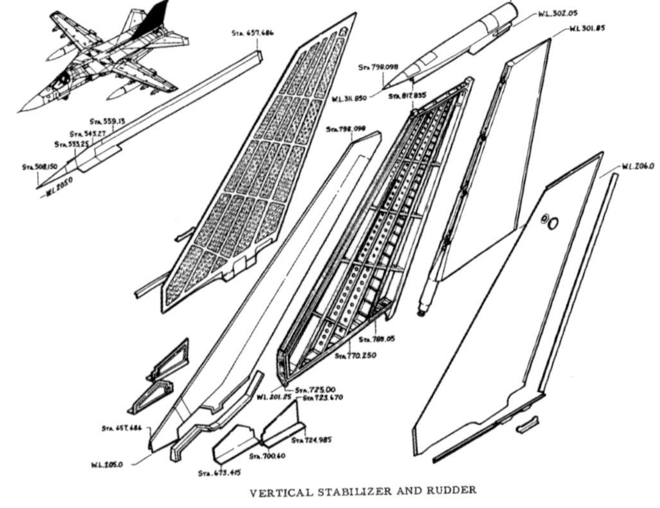

VERTICAL STABILIZER AND RUDDER

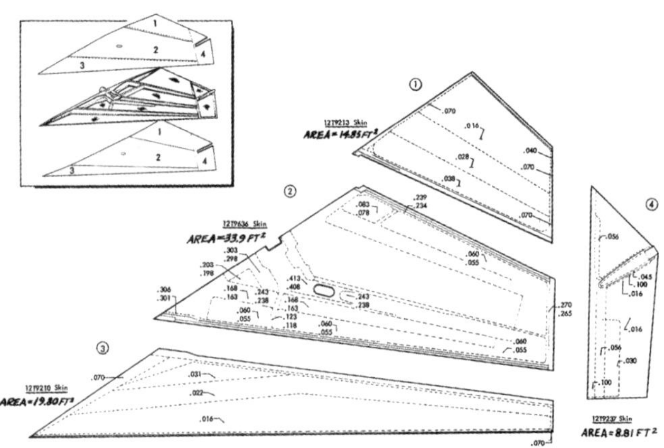

HORIZONTAL STABILIZER SKINS

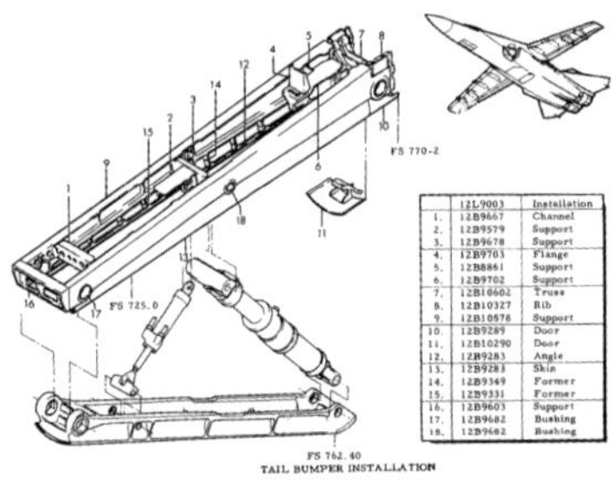

TAIL BUMPER INSTALLATION

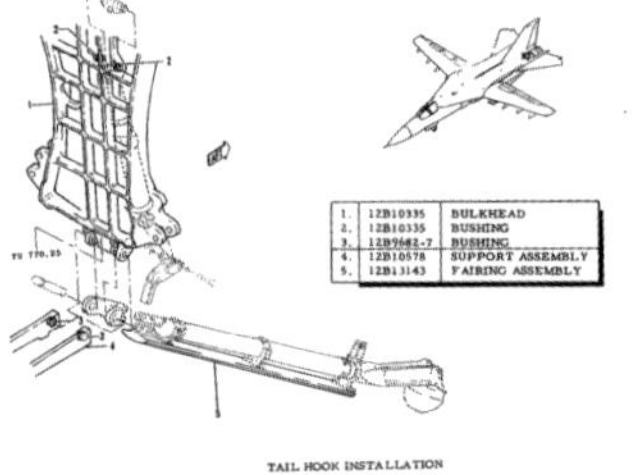

TAIL HOOK INSTALLATION

F-111 diagrams highlight the aft major assemblies. At top left is seen the inner construction of the entire aft fuselage section. Bottom left shows the parts making up the movable horizontal stabilizers. The vertical stabilizer and rudder, tail bumper, and arresting hook all are shown from top to bottom on the right. *All images this page from General Dynamics*

December 21, 1964, Fort Worth, Texas. F-111A, 63-9766, comes in to park after its first flight, with nosewheels chocked. The dashed red lines on forward section of right engine air intake show where the translationing cowl extension operates. The light-gray/white border has a scalloped wavy line as opposed to a straight one. Red stripes on the nosewheel door are ground-towing limits for turning the nosewheels. The crew escape capsule, not yet operational, was temporarily fitted with ejection seats in the first eleven early-production planes. *General Dynamics*

F-111A ship 1, 63-9766, parked in August 1967, with a BDU-6 / Mk. 43 1-megaton, free-fall, implosion-type atomic bomb designed for "lay-down" (high speed / low altitude) delivery under the port wing. BDU refers to the initialism used for "bomb dummy unit," used for training and weapons carriage/release flights instead of an actual Mk. 43 device. In a real-world scenario, the plane would streak inbound to its objective at supersonic velocity and drop the weapon, which would deploy a 23-foot-diameter parachute to slow its decent. As the bomb fell to its target, the F-111 egressed the area at high speed to escape the heat, radiation, and blast wave of the nuclear detonation. *Air Force Flight Test Center*

December 21, 1964. F-111A, 63-8766, first flight image showing slow-speed profile with leading-edge slats and aft flaps deployed. *General Dynamics*

F-111A, 65-5703, the twenty-first example fitted with a canister containing a spin chute for use if the plane experienced an unrecoverable spin situation during flight tests. *General Dynamics*

F-111A, 63-9766, with lowered main gear door, also used as a speed brake. Also seen slid forward to allow more air into the engines is the translationing intake cowl. *General Dynamics*

This view of an F-111A shows clean undersides and soot aft of the landing-gear doors from gun firing. Gun use was to be primarily for close air support. *General Dynamics*

F-111A number 2, 63-9767, topside view as it flies over a desolate landscape close to Edwards AFB in California in May 1965. Wings are at an intermediate setting. The red stripes at VG hinge location are marked at known wing sweep settings. Wing glove vanes have not yet been installed. Wing glove itself at forward end, the section above the USAF roundel, forms part of capsule aft fairing, helping it maintain stability in the slipstream once it has severed from the fuselage as a result of an ejection sequence. Flight crew wear orange coveralls, for better visibility against the desert floor in case of a search-and-rescue operation. *National Archives photo via Dennis Jenkins*

F-111A, 63-9766, catches a BAK-12 runway barrier cable in a certification test at Edwards AFB in July 1968. *National Archives via Dennis Jenkins*

F-111A, 63-9773, passing underneath a chase plane in May 1966 at Edwards AFB. Red areas exposed under slats indicate they are deployed. *National Archives via Dennis Jenkins*

The cable is attached to a thick nylon tape, wound around B-52 disc brakes, slowing the plane down to a stop. *National Archives via Dennis Jenkins*

F-111A, 63-9773, has wings swept forward. Black lines on top of the fuselage are areas where maintenance crews can safely walk. *National Archives via Dennis Jenkins*

F-111A, 63-9767, the second example built, in May 1965, with F-111A, 63-9782, *below*, the seventeenth example, in May 1967, while taxiing out for flight tests. In both pictures, the Triple Plow I (TP-1) intake translationing cowls are in their extended positions, indicated by the black fuselage slot opening forward of the main wheels. With cowls extended, this provided increased airflow volume to the engines. These function during low-speed flight and when on the airfield. Triple Plow I also features inlet spikes, which also move to control air entry, and large splitter plates close to the fuselage sides, designed to divert unstable boundary layer air away from the engine compressor sections. This was all done in attempts to eliminate compressor stalls, which plagued the program from the beginning. *Top photo General Dynamics, bottom photo Terry Panoplis*

F-111A, 63-9769, the fourth example with the F-111B wing, is test fitted with AIM-54 Phoenix missile bodies with wing movable surfaces deployed. *National Archives*

In-flight photo shows AIM-54s underwing, at low-speed test with slats/flaps/gear lowered. Date is December 1965, near Edwards AFB. *National Archives*

Another angle shows the pylons, wing glove open, the leading-edge slats lowered, and cowl slid forward. May 1964, at Edwards AFB. *National Archives*

F-111A, 63-9769, has four AIM-54s and spin chute. The aft fairings on the missile bodies are used to improve airflow during initial tests. *National Archives*

An F-111A gets weighed in a hangar. Operational weight averaged 49,100 pounds, including two aircrew of 514 pounds, engine oil (65 pounds), and 25 pounds of oxygen. *General Dynamics*

An F-111A poses with supporting crews and equipment. When planes were deployed overseas, all this hardware followed and set up shop, ready for action. *General Dynamics*

Teams of engineers gathered data in real time as planes flew. Data were also recorded and examined after a test flight ended and tapes were played back. *General Dynamics*

Gun ammunition, AIM-9 air-to-air missiles, "special weapons," and free-fall bombs of various sizes are arrayed in front of an F-111 with wings swept fully aft. *General Dynamics*

F-111s were intended to carry many external loads, from bombs to fuel tanks and racks for many munitions of different types. Several tests were carried out at multiple locations to conduct such trials. Top image shows F-111A, 63-9766, at Edwards AFB in June 1971, with a camera rack on the outboard pylon, two 600-gallon drop tanks, and a BDU-8 / Mk. 43 at the inboard station. The bottom image, also at Edwards in March 1971, shows eighteen M117 general-purpose bombs. With the loads shown here, the wings could not be moved aft since the two outboard pylons were fixed, unable to swivel into the airstream. *Both photos from National Archives via Dennis Jenkins*

An F-111A at Eglin AFB in November 1967, in between flight trials during weapons carriage/release certifications. Paint scheme is of Gull Grey FS36440 on the upper surfaces, with White FS17875 on control surfaces and undersides. A long hose provides conditioned air into the interior-compartment electronics bays to keep them cool with power on. Weapons bay has lowered racks of diagnostic equipment exposed. Open bays aft of radome contain radar avionics. One advance the F-111 made use of was the miniaturization of electric components and of microcircuit technology. This allowed for more-compact carriage of an increased number of "black boxes" capable of many more functions than in past applications. *National Archives via Dennis Jenkins*

A fine color profile of F-111A, 63-9766, at a high cruising altitude while the copilot gazes at the photo ship. In this flight profile, the translationing cowls can be retracted aft to their closed position. Angled red stripes just aft of inboard flap show known positions for flap settings. Wing is clean, with no control surfaces deployed, and no pylons attached. Solid red vertical stripe between stabilizer and strake is a warning label indicating that this area is the engine turbine wheel disintegration section of the fuselage. If an engine were to come apart while operating, this is where turbine blades are most likely to be ejected through the fuselage. *National Archives via Dennis Jenkins*

F-111A 63-9766 during roll out event, tow bar removed and a yellow hydraulic system external cart is connected to the plane to supply pressure and fluid as the wings are moved aft to the 72-degree position, one of a variety of settings that can be selected by the aircrew for high speed flight. A technician is visible in the co-pilot seat looking aft to check wing position. As seen in this view, visibility throughout the right field of view is quite good. But in this side-by-side seating design, only the right-hand occupant has the ability to scan the right sector. The same is true for the Pilot in the left-hand seat being able to cover the left sector. Aircrews adapted by being responsible for scanning their sides of the aircraft as a matter of routine, as they conducted their missions. *General Dynamics*

F-111A, 66-0011, early aft section had a fixed lower extension matching rudder, with fuel vent. The two cones were later removed. *National Archives*

When venting fuel, the vapor trail can be ignited with engine afterburners engaged. This makes for a dramatic air show scene! *General Dynamics*

F-111A, 67-0038, shows an hourglass shape below the rudder. Cylindrical housings contain clusters of ALQ-94 ECM system aerials. *Lou Drendel collection*

This F-111 lights up its vented fuel during an air show in 1983. This method also burns the fuel completely, so the mist will not land on the ground. *National Archives*

F-111s eventually adopted a two-tone and then a three-tone camouflage paint scheme, to better blend in with foliage and desert backgrounds at low level. As seen here, this does provide some blending in with the terrain underneath. F-111A, 66-0012, has AF66 on the tail fin in black, cameras facing aft on the nose and facing inboard under the wingtips, in order to record weapon drop test events in July 1969. *Air Force Flight Test Center (AFFTC)*

F-111A, 67-0061, in camouflage paint over a rugged landscape near Nellis Air Force Base, Nevada, in May 1980. Assigned unit is the 391st Tactical Fighter Training Squadron (TFTS), of the 366th Tactical Fighter Wing (TFW). MO tail code indicates that this aircraft is stationed at Mountain Home AFB, Idaho. The Tactical Air Command (TAC) badge is above the tail code letters. Red stripes aft of crew cockpit are alignment guide markings used by the tanker aerial-refueling operator during in-flight replenishment. Multiple ejector racks or MER's underwings clutch live Mk. 82 gravity bombs, six per rack. Bombs are live, as denoted by yellow stripes around forward sections where the nose fuses are inserted. Pylons swivel to maintain alignment fore/aft with the airstream as the wings are moved. *Ken Hackman / USAF*

Four F-111s make a pass over the bombing range bull's-eye, with variable-geometry (VG) wings set at different angles. The leader is flying with a 72.5-degree aft sweep. *Tony Landis collection*

This view shows the four-ship from a lead position, with all casting shadows below. Lead ship has engines to a higher rpm, compensating for decreased wing area. *Tony Landis collection*

An F-111 gets a final check before the flight crew arrives. This right-side view shows rectangular strip light, with the armament area above where munitions crews can annotate gun, bomb, and missile status. *David C. George / USAF*

This F-111D has an opened radome, exposing the APQ-130 radar dish. Below this are a pair of APQ-110 terrain-following radar antennas. One transmits and the other receives the ground returns. *David C. George / USAF*

In this artist's rendering for General Dynamics, F-111As are seen in various settings. Deployed to a "bare base" austere location, they are taking off, taxiing to depart, being maintained with wings swept back, while a pair flies overhead at high speed. All of this is being done with minimal supporting equipment. In reality, a scene like this never occurred, due to the need for an established base facility to deploy to, with personnel to take care of the many systems the plane utilized. *General Dynamics*

F-111A ship number 8, 63-9773, in May 1969 on the ramp beside an SR-71, ready for an Edwards AFB Open House event. With wings swept aft, the plane takes up a minimum of parking space. Supports underneath the horizontal stabilizers keep them level, since they tend to droop without hydraulic pressure in the system. Grey paint on topsides with an irregular borderline to white undersides. This aircraft eventually became a GF-111A, ending its days as a static instructional trainer airframe at Sheppard AFB. *Terry Panopolis via Dennis Jenkins*

RF-111A: Reconnaissance Variant

One F-111A, 63-9776, General Dynamics manufacturer's serial number (MSN) A1-11, was modified to test out the plane in the role of a reconnaissance platform. The weapons bay was adapted to load, instead of munitions, a pallet containing the tools of the trade for capturing data for intelligence value while flying in hostile airspace. A digital control system tested, monitored, and operated the cameras, an infrared (IR) line scanner, and a side-looking radar (SLR) unit. After landing, the digital suite could also ground test the equipment. This configuration was tested from December 1967 to October 1968.

Had the RF-111s been built, they would have been included as part of the overall F-111A production. Enough of them would have had the ability to provide tactical-reconnaissance sorties day and night. They were essentially the -A version, with the TFR set, increasing survivability factors of the flights since the aircrew could have used NOE (nap-of-the-Earth) flying to get in and out without being detected. Over time, inevitably more gear would likely have been used as pallet hardware became outmoded, and the fuselage itself was modified to accept more-clandestine gadgets. The idea of F-111 recon missions may have been a valid one if funding and the exotic nature of the package had not been terminated early.

The RF-111A takes off with an F-106B Delta Dart chase plane. This variant would have been aerodynamically clean, with no pylons or armament, common to most recon models. *National Archives*

RF-111A in flight. The two long, slender fairings on either side of the fuselage, below splitter plates and roundel, are the SLR antennas. A suite of cameras/sensors are in the center. *National Archives*

General Arrangement Diagram

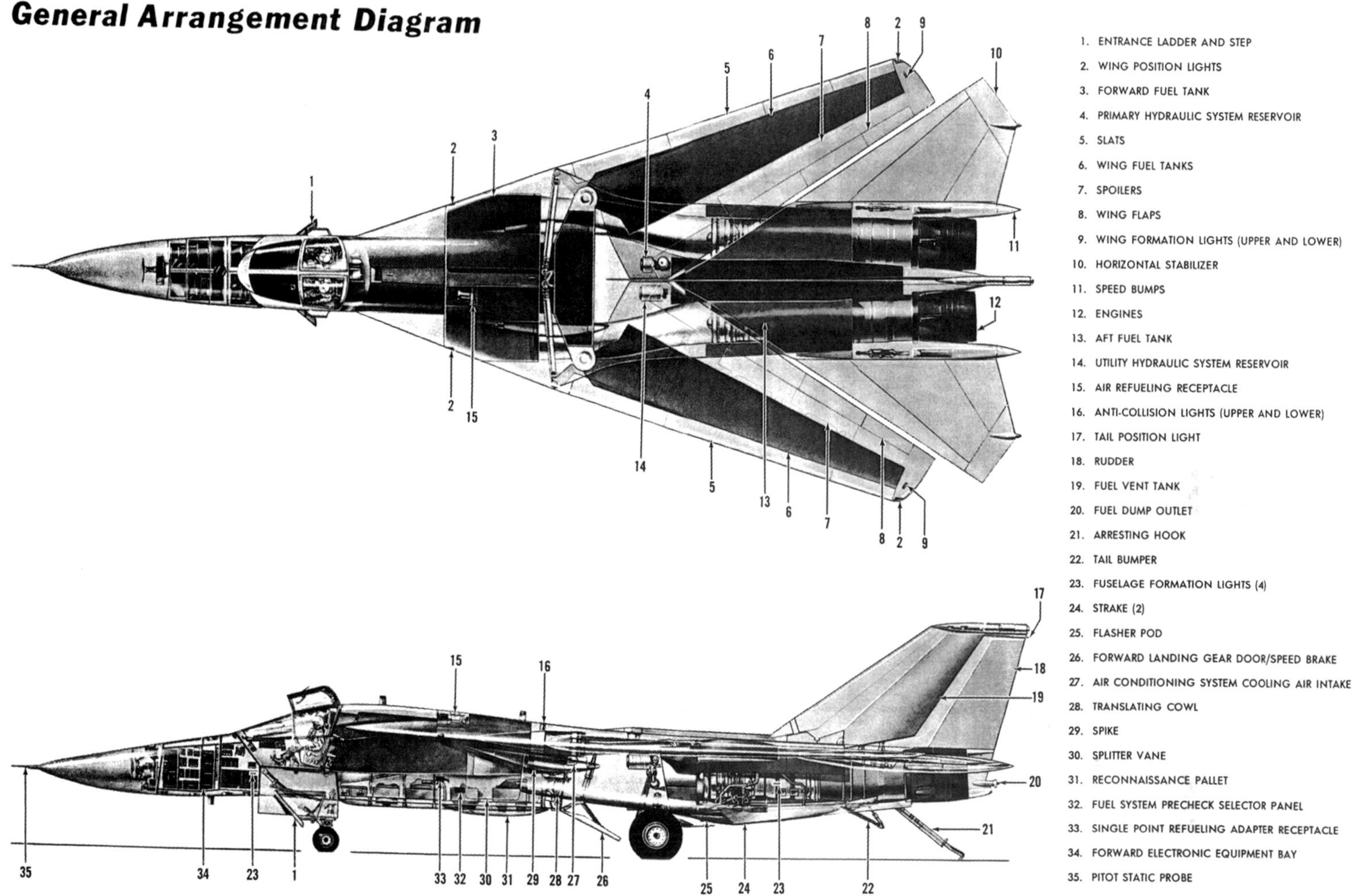

RF-111A general arrangement diagram. Entrance doors and steps (#1 on chart) are shown, as well as the layout of the reconnaissance pallet. The flasher pod (#25) was used to illuminate scenes underneath the aircraft at night for photo imaging. Aerial-refueling capability was also retained. *USAF*

An F-111B scale model is shown ready for wind tunnel testing with wings forward, four Phoenix missiles, and landing gear down, to investigate handling in low-speed configuration. *NASA*

An F-111B takes off. Cowls, wings, landing gear, empennage, and crew seating are identical to the F-111A version. Triple Plow I intakes are also a standard item. *General Dynamics*

F-111Bs, 152715, 151972, and 152714 on the ramp during testing. The Phoenix missile system emblem is on 151972. All have the infrared heat detector pod under the radome, and weapons bay doors open. The radome opened upward instead of sideways, in order to minimize the space occupied by the plane while on the aircraft carrier flight deck, or while in the hangar deck space. *General Dynamics / Grumman*

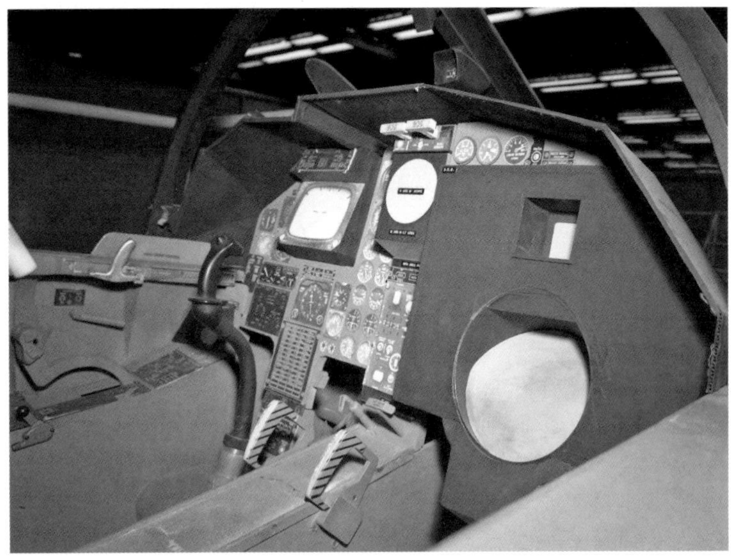

F-111B, 152714, early panel layout showing artificial horizon to the right of the altimeter. Below this is an airspeed indicator in knots, and a Mach number gauge. *Mick Roth collection*

As the F-111B layout advanced, VDIG, AWG-9, system was laid out among the other panels. The wing sweep handle above throttle was not adopted. *Grumman*

F-111B, 152714, right side showing missile control officer (MCO) station empty except for ballast plates on shelf used to simulate weight of fire control panels. *Mick Roth collection*

AVA-3 vertical-display-indicator group (VDIG) is seen here in F-111B, 151972. This was the pilot's primary instrument, used in all flight phases. *Mick Roth collection*

F-111B, 151972 VDIG. This is the direct view indicator (DVI). A projection indicator (PI) is above the glare shield and is used as a head-up display. *Mick Roth collection*

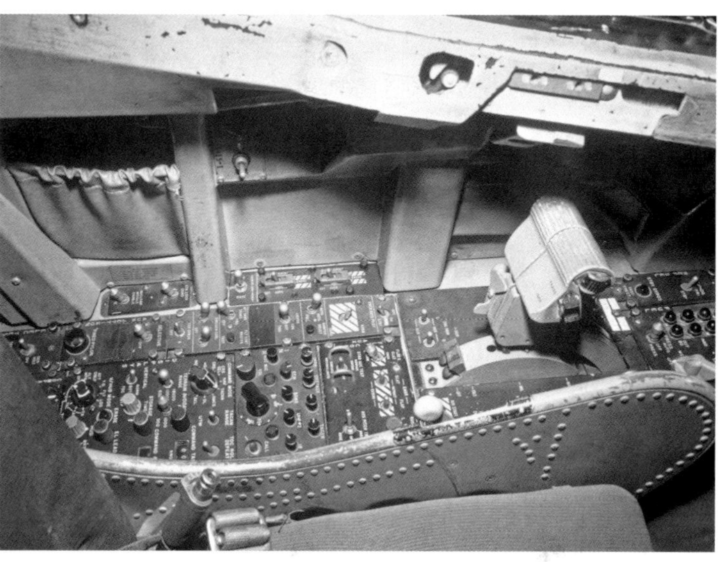

Pilot left-hand console has two panels outboard for air inlet spike operation at center. To its left is the nuclear-missile enable switch. *Mick Roth collection*

F-111B, 151972, lower-left pilot console has landing-gear handle forward. Above this is the external stores jettison switch, which is guarded against accidental operation. *Mick Roth collection*

F-111B pilot seat, canopy closed. Placards on canopy frame at right are instructions for using the radiation shields, to be placed after dropping an atomic bomb. *Mick Roth collection*

Center console between aircrew has coolant flow panel to the fire control system, low- and high-speed recorders, cameras, autopilot, and electric and ECS settings. *Both photos Mick Roth collection*

Central instrument panel has engine gauges, TACAN, hook up/down, and emergency hook-lowering handle. Below this is the main data link panel. IFF and alternate gear down at bottom.

Left: Two views of the AWG-9 system installed in F-111B, 151972. The upper image shows the detail data display panel. Thumbwheels to right of inlet cowl panel are for IR and radar manual settings. The lower-left image shows the tactical situation display scope. Above it are buttons for range settings from 25 to 400 miles ahead of the aircraft. A large LAUNCH button is at center right, below selector knob. Above is an AWG-9 system in an A-3 Skywarrior, used as a test bed for operational trials prior to installation in fleet aircraft. *All photos from Hughes*

F-111B, 151972, taxis out to conduct a flight test with an AIM-54 Phoenix missile on the port inboard pylon. Cameras under the wingtips recorded missile separation, motor ignition, and flight action. Red fins and black stripes increased visibility during testing operations. Pylon was not fixed, swiveling to maintain forward orientation as wing sweep angle changed. *Hughes*

F-111B pilot view of an XAIM-54A inert missile before takeoff. This is the #2 inboard pylon, with wing slats partially deployed. Top-right photo was taken in April 1968, as missile, plane, and fire control systems were being further integrated. Lower-right photo was taken during flight tests in October 1967, wings swept and pylons pivoted into the direction of flight. Maximum load of four missiles is shown. Infrared sensor is used to detect in a passive mode the heat signatures of hostile aircraft at long ranges, instead of using active radar to give away position. By closing on an enemy aircraft, there is less time to react once a missile is fired. *All photos from General Dynamics*

In this artist's rendering, F-111Bs are at sea in various stages of flight, from a high-speed flypast by a pair of planes at top, to a trio passing by at lower altitude at slow speed, wings forward. The USS *Enterprise* (CVAN-65) steams ahead, while one plane is on short final to trap a deck wire. Another plane has just taken to the air from the port bow catapult, while the starboard one is about to execute a launch. Two planes await behind it as one at middeck has wings swept aft, showing how compact it is while riding the deck elevator. Sadly, for some, this scene never occurred in reality. *Grumman Cradle of Aviation Museum Archives*

CHAPTER 4

F-111A in Service and Vietnam

F-111A, 66-0016, is clean and has wings at an intermediate setting. Visible along the side of the nose, just aft of the radome, is an array of radar-homing-and-warning-system (RHAWS) aerials. *General Dynamics Fort Worth Division photo*

Despite the technical hurdles being addressed amid the unrelenting stream of unfavorable press the plane was generating, F-111As rolled off the production line at Ft. Worth and were delivered to the USAF. With stated values such as ferry range (a claimed 4,180 miles versus 2,300 actual miles), a dash speed at low level of 30 miles (well short of the planned 210 miles), ongoing compressor stall issues, high drag in the aft section, steady worries about weight factors, and wing carry-through box or WCTB deficiencies, the F-111 slowly made itself known and began converting skeptics, from pilots to ground crews and those who saw the potential of everything the aircraft ultimately offered. The first base assignment for F-111As was at Nellis AFB in Nevada in July 1967, where the 4481st TFS (Tactical Fighter Squadron) was located. It was here that the Air Force had a chance to begin evaluating the plane in training missions to explore what it could do. The results were mixed. Pilots praised the ability of it to drop bombs on target as more weapons were cleared for use. The TFR capability was seen as a game changer, when it all worked as advertised. At the same time, crews were quick to criticize the side-by-side seating arrangement. With most of the F-111 pilots being from the fighter community, they naturally want unobstructed view in all directions to make best use of the "Mark-1" eyeball. In an F-111, side view for a pilot to his right is, for all practical purposes, nonexistent. The same goes for visibility of the copilot as he glances left. Eventually, crews compensated by scanning sectors of sky they could see while they flew the plane, keeping each other apprised of the outside world they could best monitor. Favorable comments centered on cockpit management, since both could work together with greater efficiency due to being next to each other, instead of not being able to directly coordinate tasks in a tandem seating layout. Another unfamiliar factor appreciated over time was that F-111 aircrews did not need to wear exposure suits, since their capsule offered them protection from the elements should they have to eject. Another negative was expressed in the desire for more maneuverability and engine power. With a misleading designation in the -F series (meaning fighter), the F-111 was, in reality, a fighter/bomber product, forever lacking the attributes commonly found in a true fighter design.

With the first five planes in place, training in earnest started under Project Combat Trident. Simultaneously, the large task of integrating it into USAF doctrine/tactics slowly expanded as flight profiles and weapons cleared for carriage increased. A further upgrade program, known as "Harvest Reaper," had a requirement to fly each plane a minimum of thirty hours in the month of September 1967. This was undertaken to silence to a dull roar those who were determined to assassinate and bury the F-111 before it ever became operational. As the month concluded, the flying time totaled up to 304 hours, double the hoped-for goal. This is even more noteworthy considering that none of the five planes were identical. Production modifications and updates were being implemented as planes were being accepted for service use. The fact that the aircraft had such a negative impression on the press, Congress, and naysayers before Harvest Reaper was attempted did not make the aircrews' lives any easier. At the squadron level, the plane grew favorably on those to the point that reenlistments at Nellis increased among the small but growing F-111 community.

F-111A production ended after 141 examples were constructed. Eighteen were of the first production batch. The last one became the FB-111A test bed. The other seventeen were RDT&E variants, with four accepted for use in 1965, eight in 1966, and five in 1967. F-111A production numbers included five in 1967, thirty-six more in 1968, eighty-six in 1969, and fourteen in 1970. In then-year dollars, the F-111A per-unit cost was $8.2 million, broken down as $4.2 million for the airframe, $1.35 million for engines, $1.69 million for the onboard electronic systems, and $925,000 for armament. The cost per flight hour was listed as $1,857 dollars.

In the background, the Vietnam War was increasing in intensity. One way to attain the coveted "proven in combat" title, as well as to tilt the balance of favor toward a weapons system, is to operate it in a theater of battle. As the F-111s at Nellis were put through their paces, a decision was made at Pentagon level, despite being a new and unfamiliar platform, to deploy the plane into the Southeast Asia (SEA) region to find out if the VG wings, TFR radar, electronic gadgets, and concepts were valid and could contribute in their own unique way toward achieving successful military objectives. Arguments for and against the deployment were based on strong factual logic. To the F-111 community, it was time to put the plane into use as it was intended to operate. That would give the opposition pause once the plane did its job in real-world conditions. On the negative side, the USAF was only into Category I–phase tests, with Cat II and Cat III yet to go. The plane was packed with a whole lot of technology not yet fully tested or judged to be reliable, so inserting it now into the crucible of combat was ill advised.

Meanwhile at Nellis, under Project Bullseye One in the spring of 1967, the F-111 was proving that it was indeed a superior platform for dropping various weapons, from gravity bombs to atomic weapons, and for firing rockets. On a flight in May to the Paris air show in France from Loring AFB in Maine, an F-111A flew there nonstop, without aerial refueling or external tanks. This progress with other shows of capability further emboldened those who desired to see the plane used in SEA. The TFR was offering up new attack options, providing more flexibility into USAF mission planning. The terrain in SEA would be an ideal proving ground for F-111 flight evaluations and refinements. In March 1968, under the combat evaluation designation "Combat Lancer," six Harvest Reaper F-111As, painted in a two-tone forest camouflage on the fuselage sides and top areas, with olive-drab-shaded undersides, deployed from Nellis to Takhli RTAFB (Royal Thai Air Force Base) in the Kingdom of Thailand.

The deployment began on March 15, 1968, with another show of ferry capability, as the planes flew across the Pacific on internal fuel, with navigation done with the inertial navigation system. All six planes (66-0017, 66-0018, 66-0020, 66-0021, 66-0022, and 66-0023) were in-country on the seventeenth with their predeployed maintenance teams and made ready for action, flying several familiarization flights in the area. On the twenty-fifth, the first F-111A combat mission was flown successfully. Typical bombloads consisted of M117 750-pound, low-drag, general-purpose weapons at wing stations #3 and #6. Two ALQ-87 ECM pods were carried, one in the weapons bay and one aft. Playing "music," as it was labeled in SEA, was jamming intended to confuse and disrupt any attempts to track or fire on an F-111 in flight. Three days later, an F-111 was lost when 66-0022 failed to return, and no concrete information existed on why the loss occurred. The flight was conducted in radio silence, at night, as a single ship flight, most of it not tracked on radar due to the low-level nature of the mission. Two days later, another plane, 66-0017 also never returned under exactly similar circumstances. With both crews lost and no explanations as to why the planes crashed, other than that the losses were evidently not caused by enemy antiaircraft weapons, the decision was made to bring the deployed numbers back up to six with a pair of planes, 66-0024 and 66-0025, flown in from Nellis and ready for operations by April 5. Flying continued until the loss of a third plane on April 22. In this incident, however, the aircrew were able to eject and were recovered. It was their debriefing that focused attention on a flight control failure as the cause of all of the crashes to date. Examination of the wreckage of this plane, BuNo 66-0024, found that a catastrophic failure of the left tailplane stabilator control unit at a faulty weld resulted in the control surface moving quickly to the limit of its travel, causing the plane to roll and pitch violently. The aircrews could not have countered the effect in time enough to recover while at low altitude and high speed. This conclusion was reinforced from clues found at yet another crash site on May 8, this one in the vicinity of Nellis. With three deployed F-111As lost in the space of a month, after flying fifty-five missions, it was decided to end the Combat Lancer operation and bring the planes home. A mixed bag of lessons were learned; the surviving planes had not been engaged by enemy AAA or detected/tracked by radar. The lost ones were not downed by hostile fire, or even by "golden BBs" as some had stated. Defective welds were the most likely culprit. To the crowd of F-111 detractors, this deployment was further proof the plane was not the right one for the tasks it was meant to perform. To add fuel to the blazing fires of negative news, it was reported that General Dynamics, in the course of ground tests, had confirmed that the vital swing-wing pivot housings were exhibiting signs of fatigue cracking. This resulted in a fleet-wide order that flight loads on the wings be limited to 3.5 g until corrective fixes could be made if defects were found during inspections. A fix was found in the form of adding reinforcing gusset plates to wing boxes, which added 500 unwanted pounds of empty weight to the plane and cost several million dollars in expense to repair. Despite this, on December 9, 1969, an F-111A on a training flight, after an attack run firing rockets was concluded, the aircrew initiated a pullout maneuver to egress the range. The entire left wing then separated from the fuselage. Sadly, the crew was lost, but it was found that the plane was not subjected to g-forces over the current restrictions in place. After the remains of the aircraft were examined in detail, it was found that a new problem was apparent. The inner swing-wing pivot section that moves back and forth against the pivot pins was found to have major flaws in its surface, causing friction resulting in cracking and loss of structural integrity.

With the F-111 program so far advanced, and with far-reaching safety implications at stake, a program was begun immediately at considerable expense to conduct exhaustive nondestructive inspections, as well as proof tests of every F-111A in USAF service. The regimen every plane had to pass involved static tests in chambers cooled down to −40 degrees Fahrenheit / −40 degrees Centigrade (a temperature at which the D6AC steel used in the wing box is most brittle and can be expected to fail), with wings set to 56 degrees sweep. The wings were then flexed over 2 meters by hydraulic rams to simulate aerodynamic stresses of from +7.33 g to a negative value of −2.4 g. Four facilities, three in Texas (two at Ft. Worth, one in Waco) and the other at the Sacramento Air Logistics Center at McClellan AFB in California, confirmed that the use of Ladish D6AC steel was a sound decision for wing box composition, and that no further issues would be expected from it if testing was passed successfully. If found to be unsafe and a replacement material was to be used instead, wing boxes have to be made from titanium and would have added a huge expense to the F-111 balance sheet. The above procedure was known as a cold-proof load test (CPLT) operation.

As these groundings, inspection programs, upgrades, and modifications transpired into the early 1970s, the F-111A eventually transformed into something that was approaching a useful tool the USAF could depend upon for low-level tactical operations. The idea of sending F-111As again to the SEA arena was floated as a way to discover if all the work done had made it an effective platform in a combat scenario. Confidence in the plane had definitely improved. A negative viewpoint was that it had done so badly, losing 50 percent of the last deployed force, that it should not be risked a second time, but the promoters of another deployment won out, with forty-eight planes of two Nellis-based squadrons, the 429th and 430th TFS, being sent back to Takhli once more, in September 1972. After thirty-three hours in-country, the six first-arriving planes were fueled up, with munitions loaded and crews briefed, and began Operation Constant Guard V as a part of Operation Linebacker I, by taxiing out and taking off on the twenty-eighth to initiate the return of the F-111A to combat missions. The rush to get the plane back into action was reminiscent of Combat Lancer. Of the six planes sent out, another one, 68-0078 (with aircrew) was lost, one was aborted due to maintenance issues, two were canceled, and two were able to perform their

F-111A at Nellis AFB. Clamshell-type canopies, avionics access, wing glove vanes rotated forward, wing slats, and flaps all seen in this view. *USAF*

assigned tasks. A five-day suspension of combat flights was ordered for crews to fly missions with the aim of getting familiar with the territory they were expected to work in, and for gradually implementing a stepped approach to low-level flights in the course of medium-altitude airstrikes accomplished first. As confidence and input was gleaned, the planes went right back to their low-altitude profiles. A common load-out was composed of stations #3 and #6 toting six Mk. 82 500-pound bombs, a single Mk. 84 1,000-pound bomb, or four SUU-30B rocket pods. The same combination could also be used on stations #4 and #5. As mentioned above, up to two ALQ-87 ECM pods could be carried in the weapons bay and on an aft pylon.

These flights were done without any external assistance from aid in the form of "Iron Hand" electronic warfare (EW) aircraft such as F-105s or EB-66s, and no support such as in-flight refueling or vectors from AWACS planes such as EC-121s. The F-111A worked totally alone—mostly at night, using approaches to targets the enemy would not be likely to monitor, and taking advantage by sneaking in through valleys and canyons and along ridgelines at high speed, the aircrew trusting the TFR systems to guide them safely through these passes en route to the blind bombing of their assigned objectives, calculated by computer and radar-generated solutions. All the aircrew had to do at the right moment was acknowledge the systems computations by consenting with the push of a button to release the ordnance loads. If used in a package of aircraft attacking targets, the F-111A found use as a "pathfinder" by directing other aircraft when to drop their munitions while simultaneously releasing their own bombs. From September 1972 through March 1973, F-111As conducted over 4,000 combat sorties, including those flown during the Linebacker II bombing offensive. Although five more planes were lost in combat (67-0066 on October 17, 67-0063 on November 8, 67-0092 on November 21, 67-0099 on December 18, and 67-0068 on December 22, 1972) and two more were lost in noncombat action (67-0072 on February 20 and 67-0111 on May 8, 1973), the mission-to-loss ratio was an admirable 0.015 percent.

After the signing of the Paris Peace Accords, F-111 basing moved from Takhli to Korat RTAFB. The move back home occurred in June 1975, but in the month of May, they were used to locate the SS *Mayaguez*, after it had been commandeered at sea by Cambodian pirates. With this homecoming, the F-111A was seen as having redeemed itself. Aircrews positively promoted the ability of the plane to carry heavy bombloads, fly under radar detection by using the natural terrain as cover, deliver weapons on targets at night and in inclement weather, and return home safely to fight another day. The variable-geometry wings, terrain-following radar, attack radar, turbofan engines, computerized systems, and navigation suite were a lot of newly integrated technology in one airframe, but it was proving over time to be a package that worked.

F-111A, 67-0072, at rest inside a hardstand at Takhli RTAFB. Load-out is twenty-four Mk. 82 500-pound, low-drag, general-purpose (LDGP) bombs. Below USAF roundel is a white ALQ-87 external ECM pod. *National Archives via Dennis Jenkins*

March 1968. F-111A, 660018, is seen after arriving to Takhli RTAFB, generating a lot of interest from base personnel. *National Archives via Dennis Jenkins*

An F-111A bombed up, secure in a hardstand with boarding ladders and a Dash-60 in place at Takhli in September 1968. *National Archives via Dennis Jenkins*

F-111A, 67-0072, overhead view shows the dorsal vents used to direct unwanted boundary layer air away from the engine intakes. *National Archives via Dennis Jenkins*

F-111A, 660018, has Mk. 82 bombs and an ALQ-87 ECM pod as it taxis out for a mission in March 1968. Tail has Combat Lancer emblem. *National Archives via Dennis Jenkins*

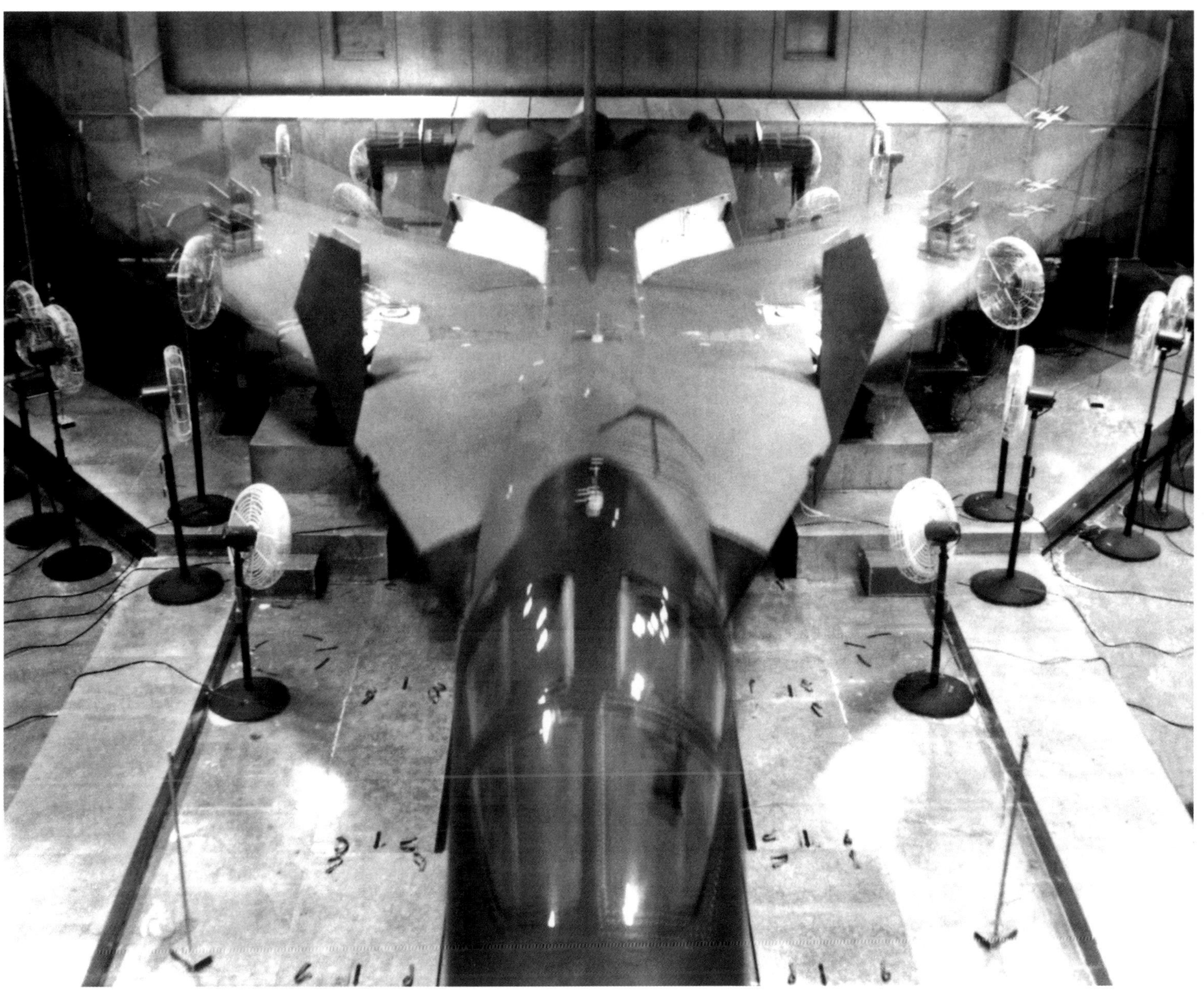

A time-exposure image while a cold-proof load test (CPLT) was carried out. Fans circulated the –40°F/–40°C air around the airframe to ensure maximum cold level required while wings were flexed repeatedly. Glove vanes have been rotated to a vertical position. In this photo the wings are set to a 56-degree sweep. Passing this evaluation proved that the Ladish D6AC steel used in the wing carry-through box (WCTB) was a good choice and did not result in a costly modification program to substitute it for another product. *General Dynamics*

This view of F-111E, 68-0007, has panels raised or removed to gain access. One tray has been slid out for inspection as a circulating air duct is connected. *USAF*

F-111A, 65-5701, takes off from Weathersfield AB, England, in May 1967. The number 199 was assigned as an aerial performer at the Paris Air Show that year. *B. Stainer / USAF*

Engine-mounted accessories were accessed by either raising or lowering aft panels. Upper right shows strake attached to panel exterior for aerodynamic purposes. *General Dynamics*

F-111A, 63-9771, with wings swept, and 63-9777 in slow flight configuration while in close formation. Radome paint differs slightly between the two. *National Archives*

CHAPTER 5

F-111C and F-111K

Both Australia and the United Kingdom expressed interest in the F-111. Foreign sales would increase production and also lower unit costs to all parties involved. It was hoped early on that several nations, mostly in the European theater, would buy the planes. The first flight in an F-111 with a pilot in command outside the United States was by Group Capt. C. H. Spurgeon of the Royal Australian Air Force (RAAF) in October 1965. Two months later, Royal Air Force (RAF) wing commander G. R. K. Fletcher operated an F-111 from the Ft. Worth, Texas, facility. In January 1966, he flew a terrain-following mission for two hours from Edwards AFB.

In the Australian case, the RAAF anticipated a buy of British BAC TSR.2 bombers, replacing the Canberras. As the TSR.2 program's development phase dragged on, the RAAF made known alternative options for a new aircraft, to include types such as the A-5 Vigilante, Mirage IVA, or the F-4 Phantom II. An eventual deal for twenty-five TSR.2s was offered, but as this unfolded, the ruling UK Labor Party was intent on scuttling TSR.2 altogether. This put the F-111 in a position as the most desirable acquisition for the RAAF. A series of talks and negotiations resulted, by October 1963, in a signed contract between the Australian minister for air, David Fairbairn, and General Dynamics for twenty-four planes, to be delivered from July to November 1968. Cost was $90,749,040 (in US dollars). At the time, the F-111 was touted as the world's best future combat aircraft, with large USAF and Navy orders in the offing. To the RAAF, the best deal they could make was now signed, sealed, and delivered; all they needed to do was wait to replace aging Canberras with frontline F-111s in a few years' time, well worth the wait. An offer before F-111 deliveries began to lease two squadrons of retired B/RB-47E Stratojet bombers from the USAF to augment the Canberras was not undertaken. Intrigued with the RF-111A modification, it was decided that six RAAF planes would be adapted after delivery to perform reconnaissance functions, with no worries about a degraded offensive capability, due to the supposed rapid change-out design of the pallet into and out of the weapons bay. The RAAF paid an RF-111A installment of $7,980,000 (in Australian dollars) as a contribution to the recon variant. The only one to fly was the single example previously mentioned.

By December 1966, the RAAF designation was F-111C/RF-111C. As 1967 began, the design included the high-lift devices / longer-span wings of the F-111B, strengthened landing gear / wheels / brakes to be incorporated into the SAC FB-111A, and a removable control stick for the copilot. A flight simulator was ordered, and USAF noncommissioned officers, with equipment, were setting up shop at RAAF Base Amberley to begin instructing RAAF crews in F-111 operations. Then the losses in SEA happened and the WCTB defects were discovered. The RAAF put the brakes on further F-111C progress until the planes were all positively inspected, with fixes found and implemented. The RAAF had every reason to be concerned. Programs to correct problems would be exponentially more expensive to fix once the planes were in Australia. The need to ensure complete structural integrity was paramount.

The first F-111C was flown in July 1968. In September, the RAAF formally accepted the F-111C into service. After the planes were flown by pilots of the manufacturer, the RAAF, and the USAF, they were partially disassembled (wings removed, avionics stored in air-conditioned spaces, fuselages consolidated at Carswell AFB in Texas, stored outside) until the fixes could be initiated. This pushed back deliveries until late 1970. Another delay resulted from the incident in which an F-111A lost a wing, as mentioned above, while one of the F-111Cs was being CPLT tested. The delays

F-111C Australian, BuNo A8-127, MSN D1-03, and USAF BuNo 67-0127, in flight. The national roundel of a red kangaroo, on a white background, with blue outer border is seen along fuselage side and top of left wing, with the three stripes next to BuNo on side of tail fin. Paint is of a three-tone SEA camouflage consisting of USAF Green / FS34102, Dark Green / FS34079, and Tan / FS30219. Undersides are of a Light Grey / FS36622. Avionics suite is same as that for the F-111A. *General Dynamics*

F-111C, A8-131, USAF BuNo 67-0131, in flight over the Nellis AFB range complex in February 2006, during exercise Red Flag 06-1, awaits a midair refueling, its flying boom receptacle open. The exceptional range of the plane is a great asset to the RAAF, operating where it does from a large landmass surrounded by the South Pacific. Blue lightning bolt on tail fin identifies this plane as belonging to the RAAF's No. 6 Squadron. A captive AIM-9 minus fins is under the left wing, and an airborne instrumentation system (AIS) pod is in shadow under the right wing. These planes also have the TF30-P-3 engines (18,500 pounds thrust with afterburning) and Triple Plow I intakes of the F-111A. Flat-gray paint scheme (USAF Gunship Grey / FS36118) seen here indicates this plane has been through the avionics update program (AUP) of the early 1990s, resulting in a software capability similar to that of the USAF F-111F version. Black radomes were eventually painted gray. *Kevin J. Gruenwald / USAF*

The first production FB-111A was this example, 67-0159. It has an early paint job with black radome. Camouflaged 600-gallon tanks hide a BDU-6 / Mk. 43 atomic bomb attached to the inner pylon. The blue diagonal stripe with white stars and the shield emblem are unique to Strategic Air Command (SAC). The entire design was often referred to back in the Cold War era as the "SAC Milky Way Sash." *USAF*

Seventy-six FB-111A models were eventually built and were ideally suited to changing SAC doctrine that dictated low-level tactics to hit strategic targets. High-altitude approaches were increasingly fraught with the likelihood that bombers would be shot down due to effective defensive measures at those heights. A low-altitude flight profile was more survivable and unpredictable and, with the proven terrain-following radar providing safe routes in and out of denied territory, provided a better chance for a successful outcome. The FB-111A was seen as filling a temporary need in SAC until the proposed AMSA (advanced manned strategic aircraft), later to be known as the B-1A, entered service.

In order to achieve the best results from attacking targets of high value, the short-range attack missile was developed. It replaced the AGM-28 Hound Dog standoff weapon. SRAM was smaller, so more could be carried, but by being small it lacked the range (100 miles / 160 km versus 700 miles / 1,100 km) of an AGM-28. B-52s could carry twenty SRAMs; an FB-111A, six (two in the weapons bay, four under the wings). A two-stage solid rocket motor propelled it to speeds of Mach 3.5. The warhead was a W-69 type with a yield of approximately 175–200 kilotons.

FB-111A, 67-0163, taxis with a load-out of AGM-69 SRAM missiles mounted on the swiveling pylons. Photocalibration markings are seen on the weapons themselves, the white outboard pylon, and the sides of the aircraft fuselage. An inertial guidance package in each missile steers it to the intended target area. Each one weighs 2,230 lbs. (1,100 kg), has a diameter of 17.5 in. (0.44 m), and has a length of 15 ft., 10 in. (4.86 m). Over 1,400 were produced and remained in the active inventory until withdrawn in 1990. *National Archives via Dennis Jenkins*

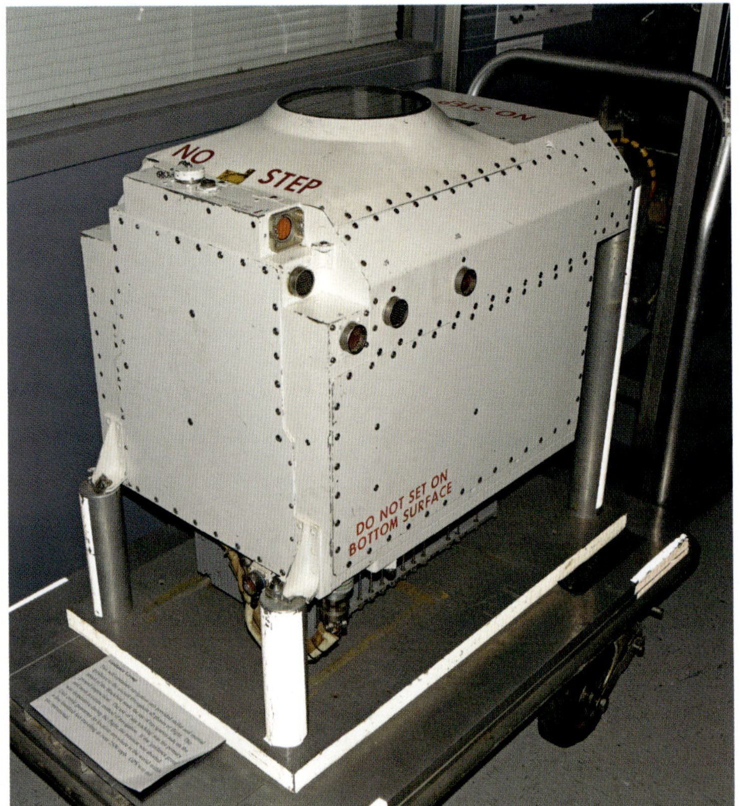

This FB-111A has an astronavigation system (ANS), seen partially in the above photo, just ahead of the black, square, automatic direction finder (ADF) antenna. Left and right photos show a similar unit, out of an SR-71, which was used to cross barren regions of the globe such as the North Pole, the primary route that bombers would fly over to attack targets in the former Soviet Union. The unit could track stars even in daylight and was accurate to several hundred feet, which was quite precise before the advent of the GPS constellation. *Above photo: USAF; photos to right: author*

FB-111A, B1-07, under construction. The USAF BuNo assigned was 67-7193. SAC emblem is above the manufacturer's serial number (MSN). *General Dynamics*

FB-111A cockpit view. Two scopes (*top center*) are for TFR (*left*) and RHAWS (*to right behind red streamer*). *Mike Henning via Dennis Jenkins*

FB-111A early trials involved the ability to refuel while in flight. This hookup is being done while carrying six 600-gallon drop tanks. Wings cannot be swept with this load-out. *General Dynamics*

An FB-111A has just released a SRAM. Cameras under the nose, wings, and aft fuselage record the event for postflight analysis. Such tests are mandatory for all combat aircraft types. *General Dynamics*

FB-111A, 68-0246, aircrew is given the chocks-out hand signal at Patrick AFB in October 1990. Triple Plow II air intake spring-loaded doors are open. *General Dynamics*

In many cases, F-111 nose art is displayed on the nose landing-gear doors. Both photos show FB-111As with the artwork of two B-24 Liberators. *David J. Ahlschwede / USAF*

FB-111A, 68-0281, in November 1977 with wings swept, pylons in line with nose of the aircraft. White undersides reflect heat from a nuclear explosion. *USAF*

FB-111As of the 509th Bomb Wing, SAC, had nose art commemorating famous bombers of the Second World War. LIBERATOR II is shown here. *David J. Ahlschwede / USAF*

CHAPTER 7

The F-111D

The F-111D variant was actually placed in service after the F-111E. The reason for this was the multitude of problems integrating the Mk. II avionics upgrade package. Included was an improved environmental-control system (ECS), radar-absorbent material (RAM) in areas of the engine air intakes, and the latest TF30-P-9 engines of 19,600 pounds of thrust each. The FB-111A landing-gear design was also incorporated. Takeoff weight was projected to be at around 100,000 pounds. The total number to be produced was planned to be a run of 315. Unit cost was $3,895,000 for the airframe, $1,229,000 for a pair of engines, $2,530,000 for the electronic fit, $6,000 for installed ordnance, and $844,000 for the associated armament. This added up to a total per-plane outlay of $8.5 million. Development costs associated with the Mk. II avionics components were estimated to be around $60 million.

The first F-111D was accepted by the USAF on June 30, 1970, with a production Mk. II suite installed. It consisted of the APQ-130 attack radar, an AYK-6 digital computer, and APN-189 Doppler navigation radar. A digital computer complex (DCC) was two units, each the size of a typical nightstand (shrunken down from the size of a refrigerator), which was a major technological leap for the time, untried in a combat aircraft until the F-111D. It did not take many test flights to discover numerous faults related primarily to the avionics systems.

The changes and promised benefits were brought on by the fast pace of advances in the electronics industry. Microelectronics enabled the creation of the well-known moniker "black box," in which many circuit boards could be plugged into a common mother board of roughly 1 square foot in area. All of this was then enclosed for protection inside a black box of aluminum or sheet-metal construction. These boxes were then connected by external plugs and wiring harnesses into a common system design. Each box provided a unique capability, such as power supply, signal generation, and so on. With the reductions in size over vacuum tube technology, less heat would be generated, and less power needed to use them. Defective boxes could be slid out of their trays, and new ones inserted easily. A new term, line replaceable unit (LRU), was coined.

Tempting as it was, the Air Force left alone the TFR set, allowing only microcircuit upgrades to increase reliability. The APQ-130, however, had improvements that could provide the aircrew with nearly photo-image-quality scenes as it mapped the ground ahead of the aircraft. It could even offer improved identification of activity below by use of an MTI (moving-target indication) mode of ground objects such as tanks or a convoy of vehicles. This was even possible in conditions of heavy rain/snow or ground clutter. If, while performing this operation, an air-to-air engagement occurred, the radar could simultaneously emit a CW (continuous wave) beam to allow missiles such as an AIM-7 Sparrow to ride this beam until it was able to engage and destroy an aerial target. All these new systems were linked together into the new digital computer and installed cockpit displays for the first time in any combat aircraft such as a horizontal situation display (HSD) and an integrated display set (IDS). All of this added up into the most advanced set of flight hardware possible in the mid-1960s. This new capability came with a few disadvantages, however: increased costs, and weeks and months of time to iron out the bugs and development expenses that climbed from the $60 million figure to nearly $300 million. To compensate, F-111D production numbers were scaled back to a total run of ninety-six examples. A hoped-for order for sixty RF-111Ds was canceled outright. All were based at Cannon AFB in New Mexico, assigned to the 522nd TFS, 523rd TFS, and 524th TFTS of the 27th Tactical Fighter Wing (TFW) under the 12th Air Force.

Being unique in hardware fit and never really fully operational, and lacking a steady supply of spare parts (resulting in low readiness rates), the F-111D was not included for upgrades later on, such as the Pave Tack modification. The constant setbacks, technological hurdles, and time spent to overcome all these issues led to a decision to build, as a stopgap measure, ninety-four examples of the F-111E variant that had some of the best attributes of the Mk. II package for less money and offered close to what had been hoped for out of the F-111D series.

F-111D, 68-0111, on the transient ramp at Patrick AFB in October 1990. The CC letters indicate the plane is based at Cannon AFB. The yellow band atop the tail fin is the color of the 524th TFTS (tactical fighter training squadron). Paint scheme is of Dark Green / FS34079, Medium Green / FS34102, Tan Brown / FS30219, and Black applied to the undersides. The USAF roundel is a black stencil on both sides. *Author*

TFR (VSD) Scope Presentations (Typical)

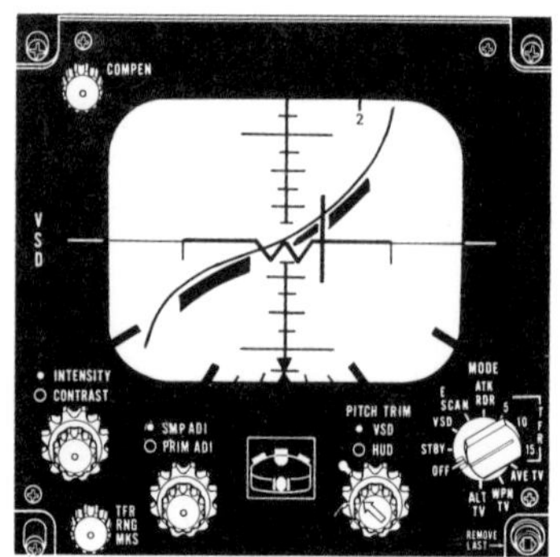

NOTE: *ONLY ONE OF THESE PULSES CAN BE DISPLAYED AT ANY GIVEN TIME.

TERRAIN FOLLOWING (TF) MODE

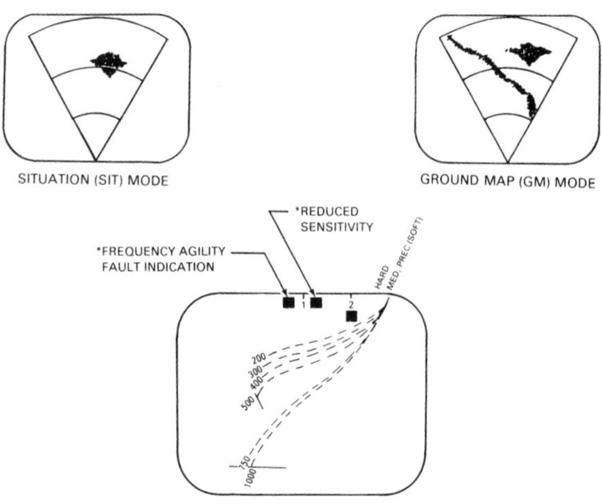

SITUATION (SIT) MODE

GROUND MAP (GM) MODE

*REDUCED SENSITIVITY

*FREQUENCY AGILITY FAULT INDICATION

HARD

MED.

PREC (SOFT)

200

300

400

500

750

1000

TEST PATTERNS

Radar Displays Approaching Mountainous Terrain Above The Aircraft Altitude

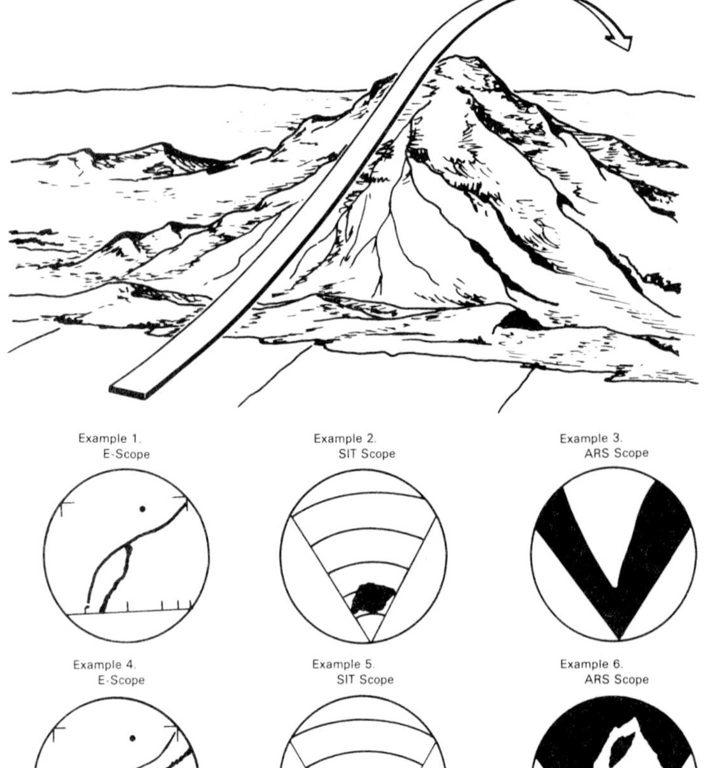

Example 1. E-Scope

Example 2. SIT Scope

Example 3. ARS Scope

Example 4. E-Scope

Example 5. SIT Scope

Example 6. ARS Scope

TFR operation as viewed on the pilot's VSD. To right is a diagram of a flight profile in rugged terrain. The term "skiing" described such low-level activity. Settings were of three choices: SOFT, MEDIUM, and HARD. HARD took you as close to the ground as possible, pulling the most g-force. Hands were kept off the control sticks while all of this was underway. *USAF*

F-111D, 68-0096, at Cannon AFB on May 12, 1992. Pylon has a Mk. / B-61 training shape. Actual weapon has a selectable yield from 100 to 300 kilotons. *Craig Kaston via Paul Minert*

F-111D, 68-0179, takes off in July 1992. Size of main gear tires, wing lift devices, wing gloves, and aft strakes all are seen clearly in this photo. *Paul Minert collection*

F-111D, 68-0151, rolls out. Flaps are set to 35 degrees. Spoilers are seen deployed. These are the first to use fly-by-wire actuators, with no mechanical backup. *USAF*

522nd TFS F-111D with a 600-gallon drop tank at Cannon in May 1979. This squadron sports red stripes on the top of the tail fin. *Paul Minert collection*

load of binders crammed full of technical data, was on hand to assist and observe. The fire chief and I walked around the capsule as it was being made ready for the upcoming test series, as he pointed out areas of interest he wanted me to photograph. After a while, the aircrew of 67-0116 on the day of the crash arrived and got into the capsule. The base photo lab had sent a team of movie techs to film the events as they occurred. When asked who was going to shoot the still images, the photo crew shrugged and stated that they had not brought a still camera, or anyone to operate one. I was immediately sought out by the GD official, and he briefed me on what I needed to shoot. He wanted me to capture the sequence of the upper panels being jettisoned and the inflation of the bags that would bring an upside-down capsule upright if it had landed in the water. I also had to shoot the location the panel was at after it had landed on the ground, in relation to the capsule itself. He said the delay would be about seventy seconds from inflation of the first bag to the start of the next panel removal / bag inflation sequence. I did as the GD rep asked, and after surrendering the film to one of the officers in charge of the overall task, we all departed back to our respective organizations. I asked the chief if there was any way I could get a copy of some of the prints, to which he assured me he would do his best, since he wanted a set as well.

A few weeks later, the chief handed me several prints, not only of the crash scene but of the work I had done during the capsule operation. It is these images that appear in this book.

In the left photo, crash crews aggressively attack the fire after the remains of 67-0116 came to rest on runway 01. Large truck in the center is an A/ S32P-2, with a capacity of 300 gallons of water and 500 gallons of AFFF foam. It has two engines, one of which drives the pump when in pump mode. Truck to the right, with radio call sign "Red-72," is an A/S32P-4. Object in foreground is one of the main gear brake assemblies. Image on right shows a cloud of carbon dioxide enveloping the burned-out fuselage as an O-6 crash vehicle expels CO_2 in order to snuff out interior fires that may be lit. Crash and fire damage to the right wing is visible. P-4 (*to right*) has roof turret raised, with handline deployed by crewmen just visible in the vapor cloud. Flaps that were torn free from the wing are also seen lying on the runway. *Both photos from USAF*

A P-4 and O-6 monitor the crash scene for any flare-up of fire while crews work to make the scene safe. One of the main gear tires is in foreground. *All photos this page from USAF*

While EOD removes underwing pylon weapon separation CADs, a handlineman remains close by to provide cover as they work in a hazardous situation.

The nose of 67-0116 is seen wedged under the inboard slat of the left wing. Avionics bays are seen, as well as fire damage to the fuselage structure.

EOD removes the two Mk. 61 training shapes that had been pylon-mounted on 67-0116. Foam coats the runway, which has been closed to air traffic.

67-0116 capsule left pitot static vent with sheared aluminum (*to right*), the result of SMDC det cord used to explosively separate this section from the plane. *All photos this page from author/USAF*

Capsule pyrotechnics tests underway. One flotation bag is inflated, waiting for the next one to begin to inflate once the panel has blown off.

Aft flotation bags operated successfully. Rocket motor primary nozzle is at bottom center. Secondary nozzle at center of image keeps capsule level after ejection.

Capsule after tests have concluded. Impact attenuation bag at bottom was used on day of crash. Round holes had blowout plugs, which softened landing on runway.

F-111E, 67-0007, lands at Aviano AB, Italy, in May 1990. In between the aft strakes is an ALQ-131 ECM pod. Under wing is an SUU-21 pod. *Sergio Gava via Paul Minert*

An F-111E, 68-0005, takes off during a USAFE exercise, Display Determination, in 1989. Aircraft is assigned to the 55th TFS, RAF Upper Heyford. *David S. Nolan / USAF*

F-111E, 67-0124, of Detachment 3, 57th FWW, clutches a pair of BDU-6 / Mk. 43 training shapes at Nellis in May 1980. *Paul Minert*

F-111E, 68-0058, in its element, clean and with wings swept aft. AD indicates it is assigned to test and evaluation duties carrying and dropping munitions. *Keith Svendsen via Paul Minert*

A1-160 / F-111E, 67-0115, the first -E variant, under construction. Many more behind it are in various stages of completion. *General Dynamics*

F-111E, 68-0055, is displayed at the Museum of Aviation at Warner-Robins, Georgia. Dash-60 electric and air start lines are connected. *Author*

An F-111E at Eglin is displayed at an air show with laser-guided bombs. The seeker heads home in on laser energy reflected off a target. *Author*

The Eglin Armament Museum has this example, F-111E, 68-0058, preserved with cluster bomb units (CBUs) as a weapon load. *Author*

CHAPTER 9

The F-111F

The most successful variant of the entire F-111 production series, the F-111F, benefited from all the effective components of the Mk. II avionics suite. It added the FB-111A/F-111D navigation and digital computers and APQ-144 radar fitted to the FB-111A and also used the F-111E stores management systems. The latest design enhancements to the wing carry-through box and landing gear were also included. Boron-epoxy composites were also fitted to save weight in fuselage construction. Most positive of all was the introduction of a new engine, the TF30-P-100. This gave the F-111F a thrust-to-weight ratio of 0.53, compared to the F-111A figure of 0.39. The engine incorporated a new low-pressure (LP) compressor section with blades able to handle increased airflow. Engine thrust had a combined total output of 50,200 pounds. More efficient combustion of exhaust gases resulted in smokeless

An F-111F of the 493rd TFS / 48th TFW with four Mk. 84 laser-guided bombs (LGBs) banks over Loch Ness, Scotland. The Pave Tack pod is lowered into the airstream, the turret optics aimed at the photo chase aircraft. *Jose Lopez / USAF*

Data Link Control Panel

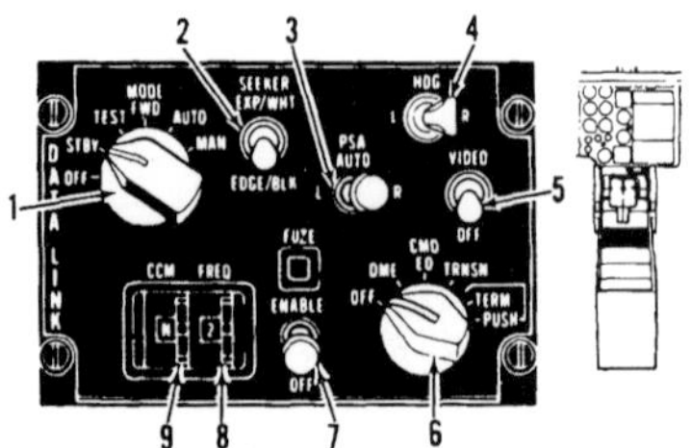

1. Mode Select Knob
2. Seeker Switch
3. Phase Scanned Array Switch
4. Heading Switch
5. Video Switch
6. Weapon Command Knob
7. Fuze Switch
8. Frequency Thumbwheel
9. CCM Thumbwheel

Laser Control Panel

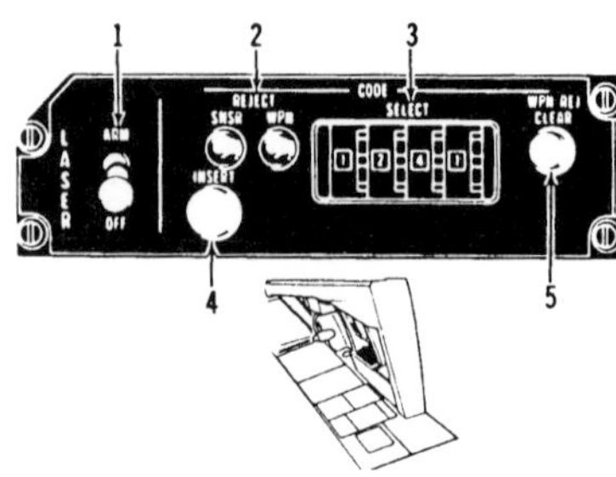

1. Laser Arm Control
2. Code Reject Indicators (2)
3. Code Select Thumbwheels
4. Insert Pushbutton
5. Weapon Reject Clear Pushbutton

Bomb Navigation Control Panels (Typical

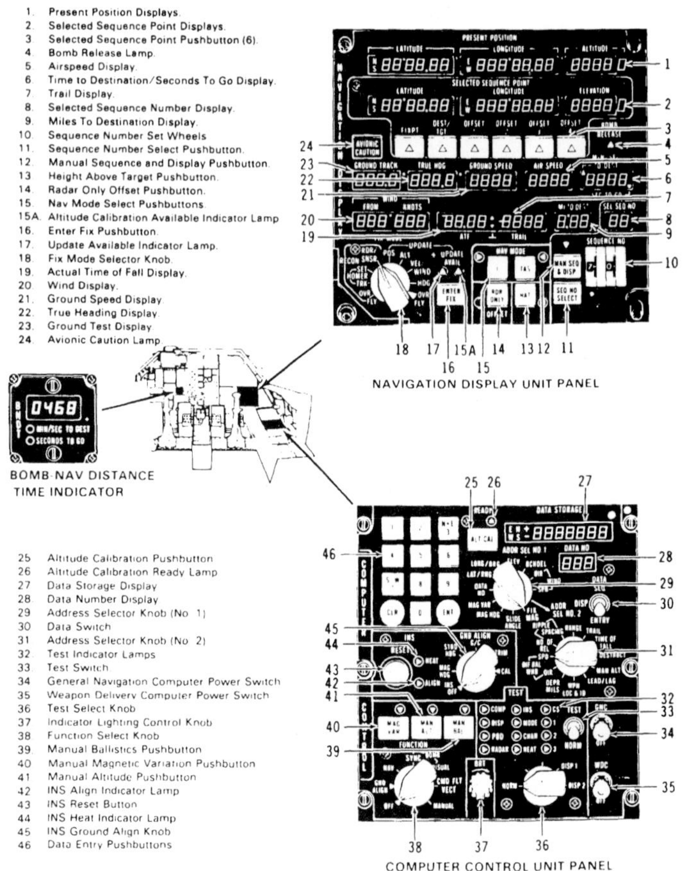

Strike Camera Control Panel (Typical)

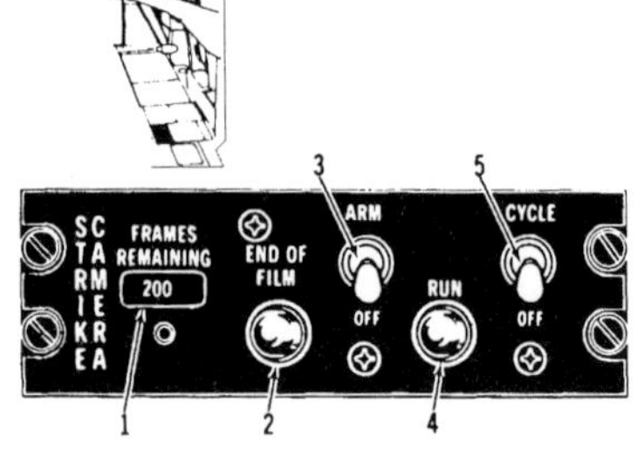

1. Frames Remaining Counter
2. End of Film Indicator Lamp
3. Arm Switch
4. Run Indicator Lamp
5. Cycle Switch

Airborne Video Tape Recorder (AVTR) Control Panel

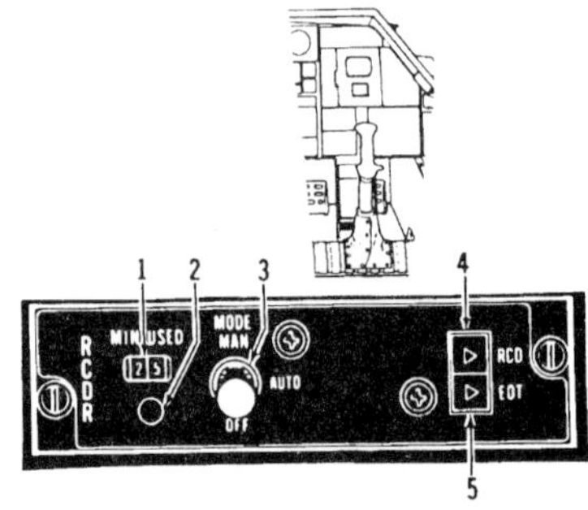

1. Minutes Used Counter
2. Minutes Used Reset
3. Mode Switch
4. Record Indicator Lamp
5. End of Tape Indicator Lamp

F-111F copilot/bombardier controlled bombing and navigation, as well as operating the strike camera for postdamage assessment study. The AVTR and laser were part of the Pave Tack pod. The data link panel connected to the AXQ-14 pod. *All illustrations from USAF*

A USAF airman in a "bunny suit" crawls inside an F-111F intake; among checks are those to ensure that the blow-in doors are unobstructed and operate smoothly. These allow additional air to enter the intakes at low speeds and while moving on the airfield. The right-side picture shows an F-111F compared to other aircraft such as the A-10A Thunderbolt II, F-15A Eagle, F-4E Phantom II, F-16A Fighting Falcon, and an F-5E Tiger II. *Both photos from David S. Nolan / USAF*

An F-111F of the 48th TFW peels away in a left turn after dropping a load of Mk. 82 high-drag bombs on the Bardenas Reales Range in Spain. In the right photo, an F-111F of the 48th TFW, after a deployment in Turkey was concluded, refuels from a KC-10A Extender over Sicily while en route back to the UK.
Both photos from David S Nolan / USAF

On a deployment to Zaragoza AB, Spain, a Pave Tack–equipped F-111F, 72-1443, releases its munitions, in this case, Mk. 82 AIR (air-inflatable retarder) bombs. The customized parachutes, known as ballutes, slow the bombs during a low-level flight profile attack so the plane can escape the blast effects. LN on the tail fin denotes this plane as being from RAF Lakenheath, home of the 48th TFW.
David S. Nolan / USAF

F-111F, 73-0707, Triple Plow II intake. Splitter plate has been removed. Probe at forward end of spike is a total Mach number indicator. *Author*

Both sides of tail fin show AF73, the fiscal year the plane was accepted for use by the government, with the last three digits of the Bureau Number. *Author*

F-111F, 73-0707, right engine air intake has five groups of vortex generators. Brown sections are sheets of RAM (radar-absorbing material). *Author*

ALE-40 chaff / flare launcher modules, located between the all-moving stabilator and the engine exhaust section. A total of 120 cartridges can be loaded on each side. *Author*

F-111F, 73-0707, left/pilot side-access hatch open. A pair of rearview mirrors are seen, a partial remedy for limited visibility aft. *Author*

F-111F pilot has primary flight gauges in his direct line of sight, essential for situational awareness in the cockpit. Light-gray panel has rows of engine gauges. *Author*

This is the view that many F-111 pilots viewed just before climbing into the plane to begin a mission. Round void space along right edge of photo is for a drink container. *Author*

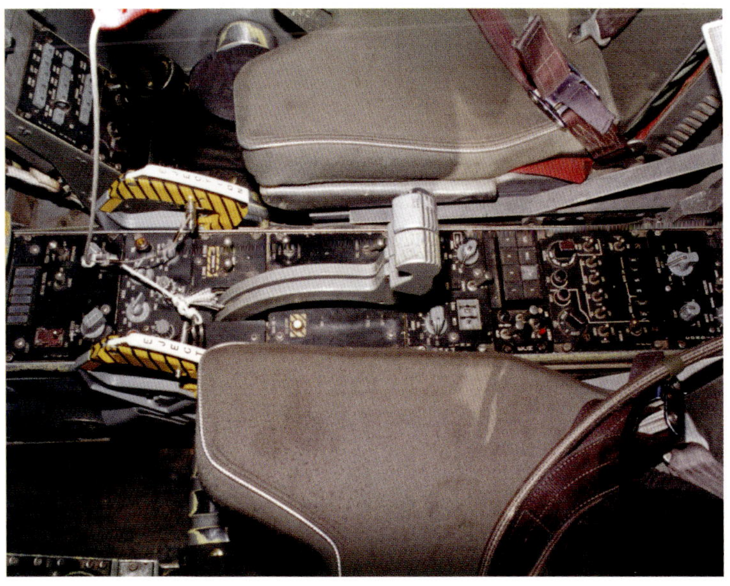

View between seats has pinned ejection handles, and a set of engine throttles the copilot can use. As throttles are moved forward, engine rpm increases. *Author*

As an F-111F approaches a KC-135E Stratotanker, the receiving aircraft checklist includes items such as turning off the radar and electronic countermeasures equipment, weapons set to safe condition, opening the refueling receptacle door, and not transmitting on the radio. When cleared by the "boomer," the plane moves in while the flying boom device hovers over the midfuselage and the probe is inserted into the receptacle. Also seen in the photo is a standby compass attached to the windscreen support and the pilot's LCOS sight to left. *David S. Nolan / USAF*

CHAPTER 10

The F-111G

F-111G, 68-0257, at Cannon AFB in New Mexico in May 1992. Blue stripe across top of tail fin indicates that this plane belonged to the 428th Fighter Squadron of the 27th Fighter Wing. Weapons bay doors are open, and under right wing is an SUU-20 pod used to drop BDU-33 or Mk. 106 practice bombs on the range. Up to four unguided 2.75-inch (70 mm) rockets could be fired from them as well, but that capability has not been used for many years. The white pods were painted the same color as the aircraft in order to further enhance low-visibility effects. Light-colored tarp is used to reflect sunlight, reducing cockpit interior temperatures. When flaps, slats, or glove vanes are used, red undersurfaces are exposed. This is a quick visual aid for ground crews to note if these movable sections are working, during control surface checks done before takeoff. Being sure that they operate is hard to do on aircraft painted completely in a dark color overall, such as on this type. *Craig Kaston via Paul Minert*

The FB-111As of SAC were intended to be in use until the B-1A became operational. When the Carter administration canceled the plane (designed with variable-geometry wings and a four-man crew escape capsule), the "FBs" stayed on as deep-strike, low-level assets. With the Reagan administration reversing the decision and ordering 100 of the B-1B Lancer version, the FB-111s were then handed over to TAC. Two squadrons were to be activated in the TAC force structure, but with the Cold War era ending and some funding outlays having to be trimmed, only a single squadron of thirty-four planes was saved, being based at Cannon AFB. The other planes headed for storage, with fifteen sold to the RAAF of Australia. The surviving examples were then redesignated as F-111Gs. Most were repainted in the Dark Grey paint scheme, losing the camouflage pattern of their former operator.

EF-111A cockpit view. The left side looks very much like most F-111 types in use, but the right side is totally devoted to managing and using the ALQ-99E, ALR-62, and ALQ-137 systems. The right-seater has no flight control stick or throttles. Center joystick is for manually slewing the navigation radar dish. Pedestal in front of large digital display indicator (DDI) screen is the DDI control interface panel. These two components are similar to what you would see inside the cockpits of an EA-6B Prowler. *USAF via Ken Neubeck*

The EF-111A conversion process was very extensive, with almost every interior component removed. Radar and avionics bays empty in this view. *Ernest H. Sealing / USAF*

An EF-111A SIR (system integrated receiver) pod is being fitted out with wiring harnesses and black boxes / antennas for the ALQ-99E ECM suite. *Ernest H. Sealing / USAF*

EF-111A conversion was done at a Grumman facility, the company already having experience on the type during the F-111B program. *Ernest H. Sealing / USAF*

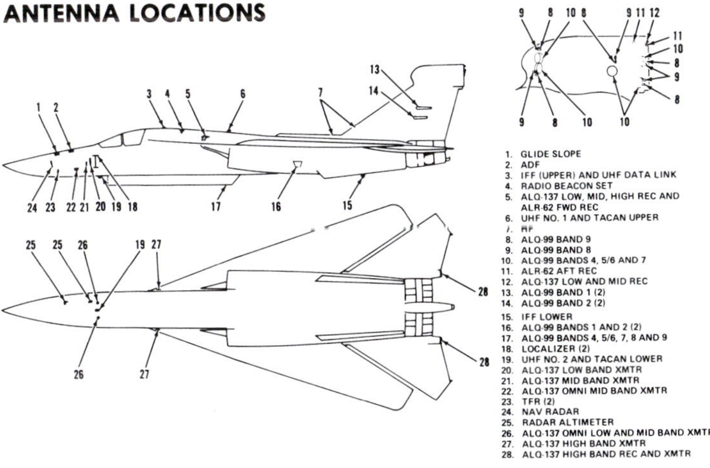

EF-111A Raven antenna locations. ALQ-99E aerials for best reception of signals were located in the fin-top fairing, with ECM and dispensers being on the fuselage itself. *USAF*

EF-111A, 66-0041, MSN EF-02, formerly MSN A1-59, in flight with a pair of 600-gallon external fuel tanks. It is claimed that four or five EF-111As were capable of blinding eastern Europe of Western air activity, from the Baltic to the Adriatic, if flying racetrack patterns over the region. Also seen here clearly is the flexible fabric seal that prevents turbulent air from entering the aft dorsal section while the wings are set to various sweep angles. The ALR-23 infrared-warning receiver optics are seen at the top of the SIR pod. They were eventually removed due to unreliability issues. *Grumman / General Dynamics*

During NATO operations in Bosnia, an EF-111A gets some maintenance work done in between missions in September 1995. *Bruce Sherwood / USAF*

Preflight walk-around of EF-111A, 66-0020, in the aft section. Exhausts, fuel dump, and ALQ-137 ECM bullet fairings are seen here. *James D. Mossman / USAF*

Nondestructive inspection (NDI) oil sample is taken from an EF-111A for analysis. Oil is checked for quality and signs for any engine wear. *Bruce Sherwood / USAF*

ALE-40 dispenser modules are documented before they are installed in an EF-111A as it is made ready for a NATO operation in Bosnia. *James D. Mossman / USAF*

An EF-111A, 68-0057, of the 42nd ECS or electronic combat squadron streaks past the Rock of Gibraltar, a strategic choke point at the entrance from the Atlantic to the Mediterranean Sea. Dark dielectric panels cover ECM, SIR pod signal receiving, and HF shunt antennas. Red stripes aid a boomer when an aerial-refueling connection is to be accomplished. The plane is participating in exercise Open Gate '89, a drill carried out with the objective of keeping the strait open to sea traffic during a simulated hostile attempt to close it. *David S. Nolan / USAF*

EF-111A, 66-0057, right midfuselage blade antenna, used for Bands 1 and 2 coverage by the ALQ-99E system. *All the photos this page taken by the author*

An aft-facing ALQ-137 ECM antenna, used for self-defense if the plane was under threat. The left engine exhaust nozzle is seen to the right.

One of a pair of ALQ-99E low/mid/high-band receivers is also part of the ALQ-62 forward sector reception array. Below this is a blue/green navigation light.

A row of four modules used to dispense chaff/flare countermeasures is seen as part of the aft fuselage, inboard of the right all-moving stabilator. Designation is ALE-40.

CHAPTER 14

F-111 Details: Weapons and Pods

F-111s, with their ability to carry large loads of munitions and supporting hardware, were as a result readily adapted to take along a wide assortment of items in flight. Bombs including free-fall, laser-guided, runway-cratering, and cluster bomb types are shown in the following pages, as are pods used for ECM tasks, radar surveillance, and data link capabilities.

The ALQ-71 used in the Vietnam era, and the lower ALQ-131, used in modern times, are shown. Both were used to provide external-jamming capabilities. *Author*

AVQ-26 Pave Tack designating pod. Turret was fully articulated to allow sensors to remain fixed on a target while the plane maneuvered. *Author*

BLU-107 Durandal runway-cratering munition in detail at the Air Force Museum. They eliminated a runway from being used to launch/recover planes. *Author*

A CBU-87 combined-effects munition is a type of CBU used to spread many smaller munitions over wide areas. The external shell opens to release the bomblets. *Author*

GBU-28 LGB gained notoriety during its baptism of fire at the conclusion of Operation Desert Storm. It weighed 4,800 pounds. *All photos this page by author*

A stores separation camera fairing on an F-111E. This one at the right-aft corner provided images of weapons compatibility with the plane for later study.

AXQ-14 data link pod, used on F-111s to relay GBU-15 images to the bombardier. He could steer the bomb by viewing real-time images, guiding it into its target.

GBU-24 Paveway III type. It weighs 2,350 pounds and has a length of 14 ft., 5 in., and a diameter of 3 ft. Glide range could exceed ten miles.

F-111E, 67-0115, with the Pave Mover radar pod. Tests with this led the way to the radar being fielded on the E-8C Joint STARS ground surveillance aircraft. *USAF*

NASA used F-111A, 63-9771, for aerodynamic research and was flown at the Dryden Research Center, in part to help cure some of the chronic performance shortcomings. *NASA*

FB-111A, 67-0159, in a distinctive red/white paint scheme, when assigned to the Sacramento Air Logistics Center, where F-111 maintenance/upgrade work was accomplished. *USAF*

A mission-adaptive wing (MAW) was flown on F-111A, 63-9778, in October 1985. The wing had smooth surfaces of flexible composite construction. *NASA*